# Cooperative Informal Geometry

Wade H. Sherard III

Dale Seymour Publications

These lessons are the result of a sabbatical project dedicated to working with middle school mathematics teachers and students in the School District of Greenville County, South Carolina. Special recognition and thanks are due to Linda W. Gunnells for her help, support, and advice in ensuring the success of the project and to Michele C. Good for her help and expertise in preparing this manuscript for publication.

Project Editor: Joan Gideon
Production Coordinator: Claire Flaherty
Art: Andrea Reider
Text and Cover Design: Lisa Raine

Published by Dale Seymour Publications, an imprint of the Alternative Publishing Group of Addison-Wesley Publishing Company.

Copyright © 1995 Dale Seymour Publications. All rights reserved. Printed in the United States of America.

Limited reproduction permission. The publisher grants permission to individual teachers who have purchased this book to reproduce the blackline masters as needed for use with their own students. Reproduction for an entire school or school district or for commercial use is prohibited.

Order Number DS21329
ISBN 0-86651-799-5

1 2 3 4 5 6 7 8 9 10-ML-98 97 96 95 94

This Book Is Printed on Recycled Paper

# CONTENTS

**INTRODUCTION** — 1

**GEOMETRIC FIGURES** — 8

| | | |
|---|---|---|
| Investigation 1 | Building Geoboard Polygons | 9 |
| Investigation 2 | Making Pattern Block Polygons | 14 |
| Investigation 3 | What Are Pentominoes? | 17 |
| Investigation 4 | Properties of Pentominoes | 22 |
| Investigation 5 | What Is a Kite? | 25 |

**BASIC GEOMETRIC CONCEPTS** — 29

| | | |
|---|---|---|
| Investigation 6 | What Is a Reflection? | 30 |
| Investigation 7 | Lines of Symmetry | 37 |
| Investigation 8 | Finding Congruent Geoboard Figures | 43 |
| Investigation 9 | Making Similar Geometric Figures | 46 |

**ANGLES OF POLYGONS** — 51

| | | |
|---|---|---|
| Investigation 10 | The Angles of a Triangle | 52 |
| Investigation 11 | The Angles of a Quadrilateral | 55 |
| Investigation 12 | The Angles of a Polygon | 58 |

**PROPERTIES OF SPECIAL POLYGONS** — 61

| | | |
|---|---|---|
| Investigation 13 | Properties of Isosceles Triangles | 62 |
| Investigation 14 | Properties of Parallelograms | 65 |
| Investigation 15 | Diagonals of Quadrilaterals | 68 |

**AREA AND PERIMETER** — 72

| | | |
|---|---|---|
| Investigation 16 | Pattern Block Areas | 73 |
| Investigation 17 | Areas of Geoboard Squares and Rectangles | 78 |
| Investigation 18 | Areas of Geoboard Right Triangles | 81 |
| Investigation 19 | Finding the Largest and Smallest Areas | 84 |
| Investigation 20 | Increasing the Perimeter | 88 |

# INTRODUCTION

*Cooperative Informal Geometry* is a collection of twenty geometry investigations designed to be taught informally using cooperative learning groups. Each investigation focuses on basic concepts or properties that are fundamental to the mathematics curriculum in geometry. Students become actively involved in the investigations, developing their understanding of geometric concepts or properties by using manipulative materials and by sharing their ideas with others. Although these investigations are designed for middle school students, some are appropriate for upper elementary school students, while others are appropriate for secondary school students who are studying geometry from an informal and intuitive point of view.

Current research in education supports the notion that learning does not occur by passive absorption alone. Learning occurs as students confront a new situation with prior knowledge, actively assimilate new information, and construct their own meanings. Accordingly, students should be actively involved in the learning of mathematics. Instruction in mathematics should engage students directly in the process of learning rather than transmit information to them passively; the use of manipulative materials and cooperative learning groups accomplishes this goal.

## MANIPULATIVE MATERIALS

Middle school students are especially responsive to hands-on activities using manipulative materials. Such activities provide them with concrete experiences from which they can construct knowledge for themselves and abstract more complex meanings and ideas. Research studies show that student achievement in mathematics can be enhanced when investigations use manipulative materials. Consequently, the investigations in this collection use manipulatives to develop understanding of geometric concepts and properties. The materials needed for these investigations include color tiles, pattern blocks, geoboards, geoboard dot paper, paper for folding, centimeter rulers, and protractors.

## COOPERATIVE LEARNING GROUPS

Cooperation is an essential part of social life; it is indispensable in the workplace, and most real mathematical work involves collaboration. Learning cooperatively in groups is an effective way to learn to work collaboratively and independently. When students participate in small groups, they compare alternative approaches, test different conjectures, share insights, express their own thoughts, state their own arguments, and learn to listen carefully to others. Cooperative learning experiences in mathematics foster improved attitudes towards mathematics and build confidence in one's ability to do mathematics. Furthermore, students in a group help each other in a low-pressure environment and learn to take responsibility for their own learning.

The cooperative learning group format provides opportunities to read, to write, and to talk about mathematics. As students talk and write about the mathematics in these investigations, they develop fundamental oral and written communication skills in mathematics by reflecting upon what they know and by organizing and clarifying their own thoughts. Students thus learn to communicate effectively their ideas, understandings, and results.

There are many different ways to structure and implement cooperative learning groups. The particular model described in the following pages is recommended for use with middle school students. The designer of this model is Eloise L. Rudy, a mathematics consultant and educator in Simpsonville, South Carolina.

### What is a cooperative learning group?

A cooperative learning group is a heterogeneous team of learners who pool their knowledge and abilities to accomplish an assigned task. A group is composed of three or four members and is structured so that each member of the team has a role with specific responsibilities. (Students working in pairs do not constitute a cooperative learning group.) The assigned task must be clearly defined with definite goals and objectives.

### What are the advantages of cooperative learning groups?

Research shows that cooperative learning groups generate interest in subject matter, foster individual accountability, promote higher achievement, develop social skills, involve all students in learning activities, put the responsibility for learning on the learner, free the instructor for more individualized instruction, encourage interdependence, allow time for informal discussion, and supplement direct instruction with more meaningful activities. Research also shows that students working in cooperative groups make significantly greater progress than students working in the traditional classroom.

### How do groups function?

Each member of the group participates in all activities. Each member is responsible for the learning of every other member. The group works on the same activity until every member of the group understands the concept or completes the activity. Assistance from the instructor is given only when no one in the group has the answer to a question.

## What are the roles of the group members?

In four-member groups, the roles are coordinator, recorder, materials manager, and resource person. In three-member groups, one person is both the materials manager and the resource person.

### Coordinator

Participates in and contributes to group activity
Gets the group settled down and started on the activity
Controls the pace of the activity and keeps group members on task
Encourages all members to contribute to discussions
Helps group members reach consensus on responses to questions
Reminds members to keep voices low during discussions
Determines when the group needs help from the instructor

### Recorder

Participates in and contributes to group activity
Keeps notes on all activities
Prepares a copy of the activity sheet to be turned in to the instructor at the end of period

### Materials Manager

Participates in and contributes to group activity
Picks up and distributes activity sheets
Picks up and distributes manipulatives in an efficient manner
Assumes the responsibility for the care of manipulatives
Collects and turns in manipulatives

### Resource Person

Participates in and contributes to group activity
Secures additional resources when needed
Asks for help from the instructor when needed
Answers questions for the group during summary activities

## What is the role of the instructor?

The instructor plays an active role in the design, implementation, and completion of the group activity.

### Before

The instructor plans the instruction for the activities to be completed. Planning includes determining the objective for the investigation, preparing any printed materials needed, and preparing manipulatives for easy distribution and collection.

### Beginning

The instructor presents the directed investigation and gives clearly defined directions for the group activity.

### During

The instructor circulates around the room, monitoring and evaluating the progress of each group. The instructor is available throughout the period to respond to questions and address problems that arise. The instructor uses a prearranged silent signal to the class when the noise level is too high.

### End
The instructor summarizes the day's activity with the entire class. With the aid of the materials managers, one report from each group is collected and all materials are returned.

### After
The instructor evaluates the group reports and assigns the same grade for the activity to each member of the group. The papers are returned for review at an appropriate time.

## GEOMETRIC CONTENT

A variety of basic concepts, definitions, properties, theorems, and formulas, all appropriate to a geometry curriculum, appear in the investigations.

### Concepts
congruence
similarity
area
perimeter
symmetry
parallel
perpendicular
bisection

### Definitions
angles: right, acute, obtuse, supplementary, complementary
triangles: right, equilateral, isosceles
quadrilaterals: general, parallelogram, rectangle, square, rhombus, trapezoid, kite
polygons: pentagon through decagon, inclusive; dodecagon, equilateral, equiangular, regular
diagonal
perpendicular bisector
reflection
line symmetry

Note that all polygons in these investigations are convex polygons. Recall that a polygon is convex if the line segment joining any pair of points in the interior of the polygon is completely contained in the interior of the polygon.

## Properties and Theorems

### Triangles

The sum of the measures of the interior angles of a triangle is 180°.

The angles opposite the congruent sides of an isosceles triangle are congruent.

Acute angles of a right triangle are complementary.

An isosceles right triangle has two 45° angles.

The bisector of the angle formed by the two congruent sides of an isosceles triangle bisects the opposite side of the triangle and is perpendicular to that side.

### Quadrilaterals

The sum of the measures of the interior angles of a (convex) quadrilateral is 360°.

The diagonals of a parallelogram bisect each other.

The diagonals of a rectangle are congruent.

The diagonals of a rhombus are perpendicular and bisect the angles at the vertices that they join.

The opposite sides of a parallelogram are congruent.

The opposite angles of a parallelogram are congruent.

The adjacent angles of a parallelogram are supplementary.

### Polygons

If two polygons are similar, then their pairs of corresponding angles are congruent and the ratios of their corresponding sides are equal.

The sum of the measures of the interior angles of a (convex) $n$-gon is $(n-2)\,180°$.

Each interior angle of a regular $n$-gon has measure $\dfrac{(n-2)\,180°}{n}$.

### Formulas

Area formulas for rectangles, squares, and triangles, and the perimeter formula for a rectangle are used.

## SUGGESTIONS FOR TEACHING THE INVESTIGATIONS

The investigations are grouped in the collection by geometric content, not by difficulty level or grade level. The investigations are independent of each other with the following exceptions: "What Are Pentominoes?" should be taught before "Properties of Pentominoes" and "The Angles of a Triangle" should be taught before "The Angles of a Quadrilateral" and "The Angles of a Polygon." The investigations vary in difficulty level, from those that are easy and direct to those that require mathematical creativity and ingenuity. They also vary in length; some can be completed in less than a class period, while others may take two class periods to complete.

Each investigation has teaching notes. These include the investigation's objective, the materials needed, the introduction and closure of the investigation, and answers to the investigation's activities. The introduction of each investigation describes the prerequisites for the lesson. The closure of each investigation briefly describes students' presentations of results and suggested facilitation in summarizing the main points of the investigation.

All investigations have been designed to be taught using cooperative learning groups. Groups of three or four students are ideal in order to promote interaction, collaboration, and the free exchange of ideas with minimal idleness or domination. Students who have had little or no experience working in cooperative learning groups will need to have training and practice in learning how to work together well. They must learn that the group "sinks or swims" together, that the group's success depends upon the effective collaboration of its members, and that each member is responsible for ensuring that the other members understand and can explain what the group has done. The group must reach consensus on the solutions to the parts of each activity and must submit one written report as a formal record of its work.

The teacher's role during the investigation is that of a manager, a monitor, and a guide. The teacher's introduction of the investigation should be brief (5 to 10 minutes). The teacher should resist the temptation to direct or lead the work of a group, to give the group answers, or to answer questions from anybody but the group's resource person. Students have been conditioned to viewing the teacher as the sole source of authority in mathematics, as the provider of correct answers to problems, and as the fountainhead of all knowledge; they must learn that *they* collectively as a group can become a source of authority by pooling their knowledge and understanding.

The degree of success with the investigation will vary from group to group. Some groups may not be able to complete the entire activity. Consequently, the closure of the investigation is critical to ensure that all students understand the essential points of the investigation. The groups should explain their answers to the different parts of the activity, and with teacher facilitation, the class should reach consensus on the solutions. The class should be led to summarize the basic points of the investigation.

If students have not had previous experiences using pattern blocks or geoboards, they will need to become familiar with them. Upon their first exposure to a new manipulative material, all students should be given an opportunity to freely explore. This is true for middle school students, just as it is for primary-aged students and adults.

## CHARACTERISTICS OF THE INVESTIGATIONS

The twenty investigations share common characteristics. These student-centered investigations
- require mathematical problem solving in discovery-learning contexts
- promote mathematical communication through having students work in cooperative learning groups
- help develop mathematical reasoning by having students collect evidence, make conjectures, and build arguments to support their positions
- make connections within mathematics by integrating several concepts in an investigation
- use manipulative materials to help students construct knowledge for themselves, develop understanding, and make abstractions
- help students develop their spatial perception and visualization skills
- cultivate social skills by helping students learn to work with others cooperatively in a group

These characteristics reflect the NCTM *Curriculum and Evaluation Standards for School Mathematics* (1989) and promote the following specific standards for middle school students:
- Mathematics as Problem Solving
- Mathematics as Communication
- Mathematics as Reasoning
- Mathematical Connections
- Geometry
- Measurement

# GEOMETRIC FIGURES

**INVESTIGATION 1**  Building Geoboard Polygons

**INVESTIGATION 2**  Making Pattern Block Polygons

**INVESTIGATION 3**  What Are Pentominoes?

**INVESTIGATION 4**  Properties of Pentominoes

**INVESTIGATION 5**  What Is a Kite?

# INVESTIGATION 1

## BUILDING GEOBOARD POLYGONS

Students will review some of the basic vocabulary of polygons, practice making geoboard polygons with emphasis on special kinds of triangles and quadrilaterals, and study the properties of these polygons. Students will also practice their spatial perception and visualization skills.

Each cooperative group is divided into two pairs. Each pair completes an activity and then checks its work with the other partnership.

### Student Materials
Each pair of students needs a geoboard and rubber bands, a centimeter ruler, and geoboard dot paper.

### Teacher Materials
Transparencies of Part A and Part B

### Introduction of the Investigation

### Part A
- Show the transparency of Part A and disclose the various sets of conditions one at a time on an overhead projector. Encourage each pair of students to check the work of the other pair in their group or to give the pair help if needed.
- Remind students that they can check for right angles by testing them with the "square corner" of a sheet of paper. Students can check for equal sides by measuring those sides to the nearest tenth of a centimeter with their centimeter ruler.
- Help students who have trouble forming parallel sides of polygons by suggesting they visualize extending those sides to determine if they will intersect.
- Remind students to check that *all* of the conditions in a set are met by their geoboard polygon.
- Show students that a rhombus can be made by first constructing its diagonals—which must be perpendicular and bisect each other—and then forming the rhombus around the two diagonals. The rhombus (in question 11) is difficult for many students to make.

GEOMETRIC FIGURES

### Part B
- Show the transparency of Part B and disclose the changes in the shape one at a time on an overhead projector for the students to make on their geoboards. Each pair of students in a group should check the work of the other pair.
- This activity uses the various special triangles and quadrilaterals that the students have made in Part A. It forces them to visualize a shape and then to decide how to transform their current shape into that shape by moving a minimum number of vertices or sides. There are different ways to do each part of this activity. Some students may find in questions 1 and 2 that they have to move more than one vertex or one side to make the next shape in the sequence. If this is the case, they should move the fewest number of vertices or sides possible.

### Closure of the Investigation
Make connections between geoboard polygons and real-world polygons by asking students to give examples of highway signs that are shaped like polygons. They should be able to describe signs shaped like rectangles, squares, equilateral triangles, regular octagons, and pentagons.

### Answers

#### Part A
1. right triangle
2. isosceles triangle
3. isosceles right triangle
4. square
5. rectangle
6. parallelogram
7. trapezoid
8. right trapezoid
9. parallelogram
10. rectangle
11. rhombus
12. quadrilateral

# GEOBOARD DOT PAPER

GEOMETRIC FIGURES

# INVESTIGATION 1

## BUILDING GEOBOARD POLYGONS

### Part A

### Geoboard Shapes

Make a geoboard shape that fits the given conditions. Each pair of students in your group will check the shape of the other pair. You may use rulers to measure the sides of shapes if you need to. Draw your geoboard shape on geoboard dot paper and identify it.

1. three sides
   one right angle

2. three sides
   two sides equal

3. three sides
   one right angle
   two sides equal

4. four sides
   all sides equal
   four right angles

5. four sides
   opposite sides equal
   four right angles

6. four sides
   opposite sides parallel
   no right angles

7. four sides
   exactly two sides parallel

8. four sides
   exactly two consecutive right angles

9. four sides
   opposite sides equal
   no sides perpendicular

10. four sides
    opposite sides parallel
    adjacent sides perpendicular

11. four sides
    all sides equal
    no sides perpendicular

12. four sides
    no sides parallel
    no sides perpendicular

## Part B

### "Change-A-Shape"

1. Make this rectangle on your geoboard. By moving only one vertex or one side, change the shape on your geoboard to the next named shape.

    - Change the rectangle to a right triangle.
    - Change the right triangle to a trapezoid.
    - Change the trapezoid to a parallelogram.
    - Change the parallelogram to an isosceles right triangle.
    - Change the isosceles right triangle to a square.

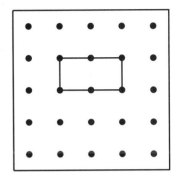

2. Make this square on your geoboard. By moving only one vertex or one side, change the shape on your geoboard to the next named shape.

    - Change the square to an isosceles triangle.
    - Change the isosceles triangle to a parallelogram that is not a rectangle.
    - Change the parallelogram to a trapezoid.
    - Change the trapezoid to a rectangle.
    - Change the rectangle to a quadrilateral with no sides parallel.

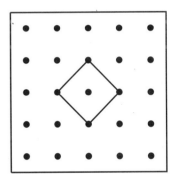

3. Make this parallelogram on your geoboard. By moving only one or two vertices, change the shape on your geoboard to the next named shape.

    - Change the parallelogram to a rectangle that is not a square.
    - Change the rectangle to a square.
    - Change the square to a rhombus that is not a square.
    - Change the rhombus to a trapezoid.
    - Change the trapezoid to a parallelogram that is not a rectangle.

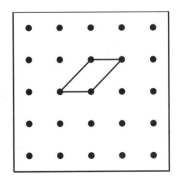

GEOMETRIC FIGURES

# INVESTIGATION 2

## MAKING PATTERN BLOCK POLYGONS

This investigation requires students to use several concepts concerning polygons with more than four sides: basic vocabulary, perimeter, and angle measure. Students also practice their spatial perception and visualization skills. Students should have had previous experience in working with pattern blocks and should know the measures of the angles for each pattern block piece.

### Student Materials
Each group needs a set of pattern blocks. The blocks on page 92 can be copied on heavy paper, laminated, and cut out for teacher-made pattern blocks.

### Introduction of the Investigation
- Review basic vocabulary of polygons, especially the terms equilateral and equiangular.
- Review the pattern block pieces and the measures of each of their angles.
- Demonstrate one of the activities for the class.

### Closure of the Investigation
The groups report their findings. They may demonstrate their pattern block shapes by making them on an overhead projector.

### Possible Answers

1. perimeter: 5
   angles: 60°, 150°, 90°, 90°, 150°

2. perimeter: 7
   angles: 60°, 120°, 120°, 120° 120°

3. perimeter: 8
   angles: each is 120°

4. perimeter: 6
   angles: 150°, 60°, 150°, 150°, 60°, 150°

5. perimeter: 7
   angles: 120°, 150°, 90°, 90°, 150°, 120°

COOPERATIVE INFORMAL GEOMETRY

6. perimeter: 7
   angles: 60°, 150°, 150°, 120°, 90°, 150°

7. perimeter: 7
   angles: 150°, 90°, 150°, 120°, 150°, 90°, 150°

8. perimeter: 8
   angles: 120°, 150°, 150°, 120°, 120°, 150°, 150°, 120°

9. perimeter: 9
   angles: 150°, 150°, 90°, 150°, 150°, 90°, 150°, 150°

10. perimeter: 9
    angles: 150°, 120°, 150°, 150°, 120°, 150°, 150°, 120°, 150°

11. perimeter: 10
    angles: 150°, 120°, 120°, 150°, 150°, 150°, 150°, 120°, 120°, 150°

12. perimeter: 12
    angles: each is 150°

GEOMETRIC FIGURES

# INVESTIGATION 2

## MAKING PATTERN BLOCK POLYGONS

You need a set of pattern blocks.
   Use pattern blocks to make polygons having the given properties. There may be more than one polygon that has the given properties. For each of your pattern block polygons
- trace the pattern block polygon on paper
- find its perimeter
- find the measure of each of its angles

1. A pentagon that is equilateral but not equiangular.
2. A pentagon that is not equilateral and has only one line of symmetry.
3. A hexagon that is equiangular but is not equilateral.
4. A hexagon that is equilateral but is not equiangular.
5. A hexagon that is neither equilateral nor equiangular.
6. A hexagon that has only one right angle.
7. A heptagon that is equilateral but not equiangular.
8. An octagon that is equilateral but not equiangular.
9. An octagon that is neither equilateral nor equiangular.
10. A nonagon that is equilateral but not equiangular.
11. A decagon that is equilateral but not equiangular.
12. A regular dodecagon.

# INVESTIGATION 3

## WHAT ARE PENTOMINOES?

This discovery-learning, problem-solving activity helps students to develop spatial perception. Students will experience several geometric concepts, especially congruence and geometric transformations (flips, turns, and slides).

### Student Materials
Each student needs five color tiles, a pair of scissors, and a sheet of one-inch squared paper. Puzzle grid sheets should be distributed *only* *after* a group has discovered all twelve different possible pentominoes.

### Introduction of the Investigation
- Discuss how to form a domino with two color tiles and a tromino with three color tiles and how many different dominoes and trominoes are possible. (There is only one way to form a domino out of two tiles and only two ways to form a tromino out of three tiles.)
- Discuss how to form a pentomino and how to decide if two pentomino shapes are the same (congruent).
- Outline the different parts of the activity. Do not tell your students the number of unique pentominoes. Distribute the puzzle grid sheet only *after* a group has discovered all twelve different pentominoes.

### Closure of the Investigation
1. Show the twelve different possible pentominoes. Discuss the following systematic way of finding them:

   | Maximum Number of Squares in a Row | Number of Different Such Pentominoes |
   | --- | --- |
   | 5 | 1 |
   | 4 | 2 |
   | 3 | 8 |
   | 2 | 1 |

2. The puzzle grid sheet has an area of sixty squares. Since each pentomino has an area of five squares, there are 60 ÷ 5 = 12 pentominoes. Show a solution to this challenging puzzle.

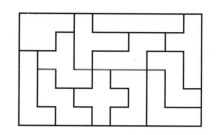

3. Challenge the students to find as many different solutions to the puzzle as they can. Post on a wall each different solution that the students discover.

**Answers**

Dominoes (two squares):

Trominoes (three squares):

Pentominoes (five squares):

**ONE-INCH SQUARE GRID**

**PUZZLE GRID**

# INVESTIGATION 3

## WHAT ARE PENTOMINOES?

You need five color tiles, a sheet of squared paper, and a pair of scissors for each student in your group; one puzzle grid sheet per group will be distributed later.

1. Make a shape with five square tiles. Make sure that at least one full side of a tile touches one full side of another tile. A shape with five square tiles is called a *pentomino*.

   This is a pentomino.

   This is not a pentomino.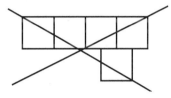

2. Using color tiles, find all the different pentominoes that are possible. Draw them on squared paper and cut them out. If a pentomino can be flipped or turned to fit on another pentomino, it is not a different pentomino.

3. How many different pentominoes are possible? Explain how you arrived at your answer.

4. Obtain a puzzle grid sheet from your teacher. Using a complete set of pentominoes and the puzzle grid sheet, fit all the pieces together to make a rectangle with no holes in it and with no overlapping pieces. Find as many ways of doing this as you can.

# INVESTIGATION 4

## PROPERTIES OF PENTOMINOES

This discovery-learning activity gives students an opportunity to confront the difference between area and perimeter, to explore the concept of line symmetry, and to exercise their spatial perception and visualization skills. Students should have completed Investigation 3 as a prerequisite.

### Student Materials
Each student needs a complete set of paper pentominoes. Each group needs at least three puzzle grid sheets made from one-inch squared paper having dimensions 12 by 5, 15 by 4, and 20 by 3.

### Introduction of the Investigation
- Review what a pentomino is.
- Explain the activity.

### Closure of the Investigation
In Part A, point out that geometric shapes can have the same area but different perimeters.

In Parts B and C, ask students to demonstrate their solutions by folding their pentominoes.

In Part D, highlight the fact that the total area of the twelve pentominoes is $12 \times 5 = 60$. Ask the students to express 60 as the product of two different integer factors in as many ways as they can: $1 \times 60$, $2 \times 30$, $3 \times 20$, $4 \times 15$, $5 \times 12$, and $6 \times 10$. Which of these factorizations can be the dimensions of rectangular pentomino puzzles and which cannot?

# Answers

## Part A

This pentomino has a perimeter of 10 units. All other pentominoes have a perimeter of 12 units.

## Part B

These pentominoes have lines of symmetry.

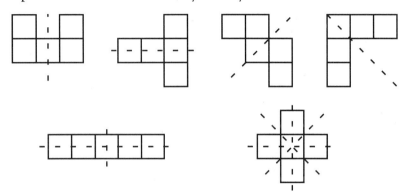

## Part C

These pentominoes will not fold up into an open box.

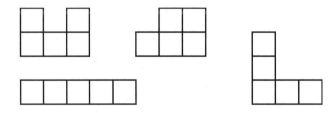

## Part D

1. Possible solutions to the rectangular pentomino puzzles

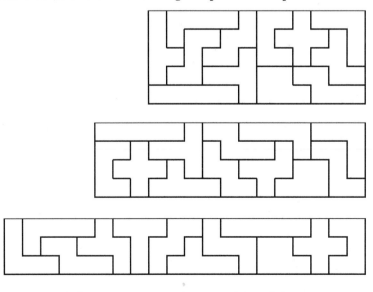

2. Since some of the pentominoes have widths of three units, it is impossible to fit them into a rectangle with a width of only two units or only one unit.

GEOMETRIC FIGURES

# INVESTIGATION 4

## PROPERTIES OF PENTOMINOES

You need a complete set of paper pentominoes for each group member.

### Part A
Each pentomino has an area of five square units.

1. Find the perimeter of each different pentomino. Do they all have the same perimeter?

2. If two geometric shapes have the same area, do they have the same perimeter?

### Part B
Determine which pentominoes have lines of symmetry. Some of them have more than one line of symmetry. Sort the pentominoes into sets that have no lines of symmetry, one line of symmetry, two lines of symmetry, and four lines of symmetry. Check by folding the pentomino along its line, or lines, of symmetry.

### Part C
Predict which pentominoes will fold up into an open box. Mark an X in the square that you think will be the bottom of the box. Then check by trying to fold the pentomino into an open box.

### Part D
1. Find different ways to fit all twelve pentominoes into a rectangle with no holes in it and with no overlapping pieces so that the rectangle is
    a. twelve units long and five units wide
    b. fifteen units long and four units wide
    c. twenty units long and three units wide

    Sketch your solutions on the back.

2. Why is it *not* possible to fit all twelve pentominoes into a rectangle with no holes in it and with no overlapping pieces so that the rectangle is
    a. thirty units long and two units wide?
    b. sixty units long and one unit wide?

# INVESTIGATION 5

## WHAT IS A KITE?

This discovery-learning activity is designed to help students to discern the common properties of a class of geometric shapes and to write a comprehensive definition for that class of shapes. Geometric concepts involved include congruence, line symmetry, perpendicular bisector, and angle bisector.

**Student Materials**
Each student needs a copy of the sheet of geoboard kites, a centimeter ruler, and a protractor.

**Introduction of the Investigation**
- Explain procedures.
- Mention that in the collection of kites some may have aeronautical designs that are better than others. Air worthiness is not an issue here. It is assumed that a kite is a convex quadrilateral in this investigation.

**Closure of the Investigation**
The groups report their discoveries about the properties of a kite. Each group reads its definition, and the class discusses the accuracy and the quality of the definitions.

**Answers**

Part A
1. A kite has two pairs of congruent adjacent sides. Its pairs of congruent sides have different lengths. It has no pairs of parallel sides.
2. A kite has a pair of opposite angles that are congruent and a pair of opposite angles that are not congruent.
3. A kite has one line of symmetry—the diagonal joining the vertices of the two noncongruent angles.
4. The diagonals of a kite are perpendicular. Only one diagonal, which can be called the *principal diagonal,* is a line of symmetry. The principal diagonal bisects the other diagonal. The principal diagonal bisects the angles that it joins.

### Part B

The definition of a kite should be as concise as possible, but it should accurately distinguish between quadrilaterals that are kites and those that are not kites. The following are possible definitions of a kite:

A kite is a quadrilateral having two pairs of congruent adjacent sides with no side common to both pairs.

A kite is a quadrilateral with two pairs of congruent sides and no pairs of parallel sides.

A kite is a quadrilateral whose only line of symmetry is a diagonal.

There is no universal definition of a kite. In this investigation it is assumed that a kite is a convex quadrilateral and that squares and rhombuses are not kites.

### Extensions

1. Could a rhombus be a kite? Could a square be a kite? How would the definition of a kite have to be modified to include these two shapes?
2. Develop a formula for the area of a kite in terms of the lengths of its diagonals. (Its area is one-half the product of the lengths of its diagonals.)

# GEOBOARD KITES

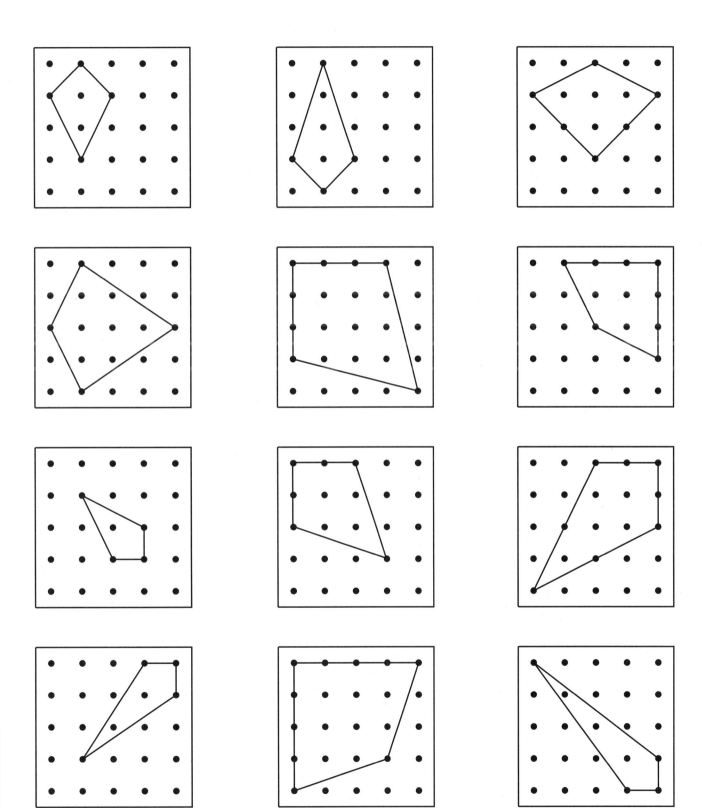

# INVESTIGATION 5

## WHAT IS A KITE?

You need a sheet of geoboard kites of different sizes and shapes, a centimeter ruler, and a protractor.

### Part A
Draw the diagonals of each kite in your collection. Then, measure its sides, its angles, and its diagonals.

1. What is true about the sides of a kite?

2. What is true about the angles of a kite?

3. Does a kite have any lines of symmetry?

4. Find all the different properties of the diagonals of a kite that you can. (How are the diagonals related to each other? How are the diagonals related to the kite itself? How are the diagonals related to the angles of the kite?)

### Part B
Write a definition for a kite. Your definition should be brief, but it should enable you to sort a collection of quadrilaterals into kites and those that are not kites.

# BASIC GEOMETRIC CONCEPTS

**INVESTIGATION 6**　　What Is a Reflection?

**INVESTIGATION 7**　　Lines of Symmetry

**INVESTIGATION 8**　　Finding Congruent Geoboard Figures

**INVESTIGATION 9**　　Making Similar Geometric Figures

# INVESTIGATION 6

## WHAT IS A REFLECTION?

This discovery-learning activity helps students to develop an understanding of the meaning of a reflection. Students will exercise their spatial perception and visualization skills.

### Student Materials
Each group needs a set of pattern blocks and centimeter rulers.

### Introduction of the Investigation
- Explain the activity, and discuss any new vocabulary such as reflection, mirror image, reflection line, and flip. If mirrors are available, each group can check reflection images by placing a mirror on the reflection line and comparing the actual mirror image to its reflection pattern-block image.
- Remind students that when they measure the distance between a point and the reflection line they are measuring along a line that is *perpendicular* to the reflection line.

### Closure of the Investigation
The groups report their solutions to the activity. They may demonstrate their pattern block reflections by recreating them on an overhead projector. Part C is especially important to discuss, since it describes mathematically the relationship between a figure and its reflection image. Conclude the investigation by developing the understanding that a reflection preserves congruence.

### Answers

#### Part C

$P$ and $P'$ are equidistant from line $m$. (Read $P'$ as "P prime.") The reflection line $m$ is the perpendicular bisector of line segment $\overline{PP'}$, for any point $P$ on the figure and its reflection point $P'$.

**Part D**

Choose a vertex $P$ of the hexagon and its image point $P'$ on the reflection image. Draw $\overline{PP'}$ and fold the paper over so that $P$ and $P'$ coincide. Pinch and crease to fold the reflection line. Choosing two pairs of points $P$, $P'$ and $Q$, $Q'$ may help to make the fold more accurate.

The folded line is the perpendicular bisector of $\overline{PP'}$ and thus is the reflection line.

BASIC GEOMETRIC CONCEPTS

# INVESTIGATION 6

## WHAT IS A REFLECTION?

You need a set of pattern blocks and centimeter rulers.

### Part A
A *reflection* of a figure in a line is a figure that is the mirror image of the figure in the line. Use pattern blocks to form the reflection image of each figure in the given line. Then trace around the blocks to draw the reflection image of the figure.

1.
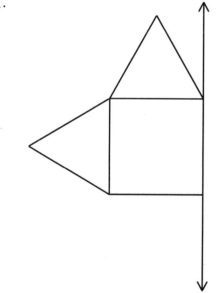

2.
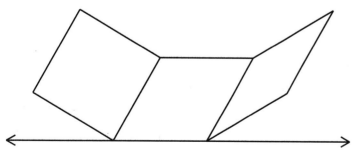

3.

4.

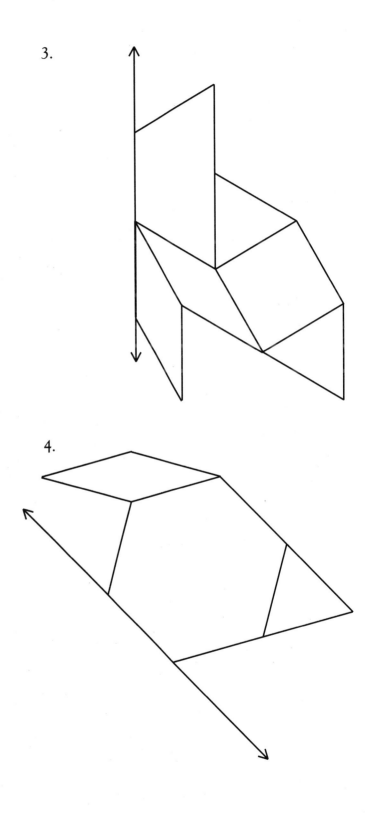

5.

6.

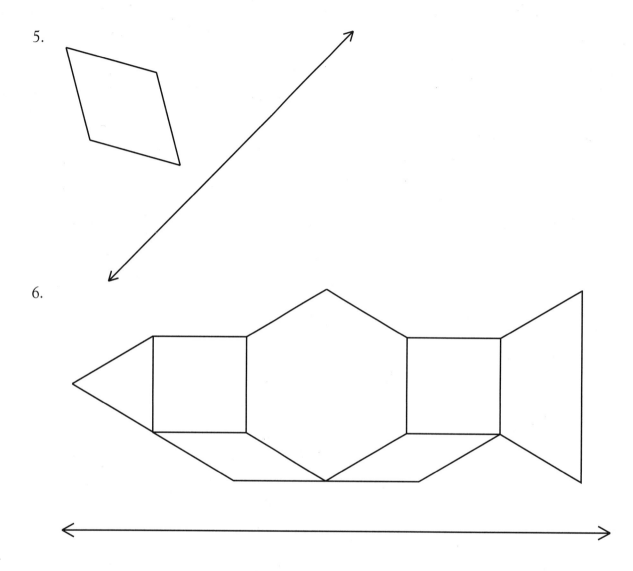

**Part B**

A reflection of a figure in a line can also be thought of as a flip of the figure in the line. Both the figure and its reflection image are flips of each other in the reflection line.

Trapezoid A´B´C´D´ is the reflection image of trapezoid ABCD in line m. (Read A´ as "A prime.") Fold this sheet of paper along line m. Check to see if each trapezoid flips on top of the other and is the reflection, or flip, of the other.

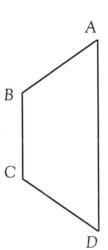
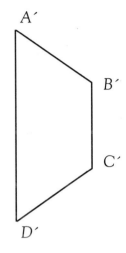

1. Draw line segment $\overline{AA'}$. Measure $\overline{AA'}$; then measure the distances of both A and A´ from line m.

2. Draw line segment $\overline{BB'}$. Measure $\overline{BB'}$; then measure the distances of both B and B´ from line m.

3. Draw line segment $\overline{CC'}$. Measure $\overline{CC'}$; then measure the distances of both C and C´ from line m.

4. Draw line segment $\overline{DD'}$. Measure $\overline{DD'}$; then measure the distances of both D and D´ from line m.

5. Mark any point along line segment $\overline{AB}$ of trapezoid ABCD and label it X. Fold the paper over along line m to find the reflection of point X on trapezoid A´B´C´D´. Label the reflection point X´.

   Now draw line segment $\overline{XX'}$. Measure $\overline{XX'}$; then measure the distances of both X and X´ from line m.

6. Mark any point along line segment $\overline{CD}$ of trapezoid ABCD and label it Y. Fold the paper over along line m to find the reflection of point Y on trapezoid A´B´C´D´. Label the reflection point Y´.

   Now draw line segment $\overline{YY'}$. Measure $\overline{YY'}$; then measure the distances of both Y and Y´ from line m.

BASIC GEOMETRIC CONCEPTS

**Part C**

Describe how a point *P* on a figure and its reflection point *P′* on the figure's reflection image are related to each other and to the reflection line *m*.

**Part D**

The hexagons below are reflections of each other. Find their reflection line. Explain how you did this and why you know that your line must be their reflection line.

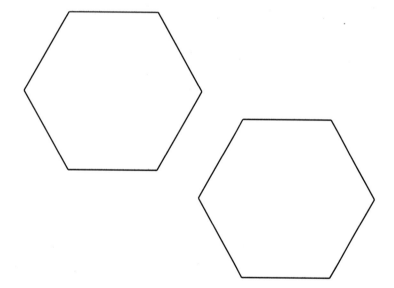

# INVESTIGATION 7

## LINES OF SYMMETRY

This investigation requires students to use their understanding of the geometric concept of line symmetry. Students will have ample opportunities to practice their spatial perception and visualization skills while exercising their creativity.

### Student Materials
Each group needs a set of pattern blocks, paper, and pencils.

### Introduction of the Investigation
- Discuss the meaning of lines of symmetry of a geometric figure.
- Talk about the lines of symmetry of familiar geometric figures such as equilateral triangles, squares, rectangles, rhombuses, isosceles trapezoids, and regular hexagons.
- Outline the activity, reminding the groups that some activities may have more than one solution.
- You may choose to discuss Part A or Part B with the entire class as sample activities.

### Closure of the Investigation
The groups present their results; they may want to use an overhead projector. Since some of the activities have different possible solutions, compare and contrast the different solutions that the groups generate. Discuss any strategies that the groups used. For example, the required pattern blocks can sometimes be used to make a shape having known lines of symmetry. If mirrors are available, the groups can check the lines of symmetry of their figures by placing a mirror on the conjectured lines of symmetry.

### Possible Answers

#### Part A

1.    2a.    2b.

2c.    3.    4.

BASIC GEOMETRIC CONCEPTS

**Part B**

1.    2a.    2b.

3a.    3b. 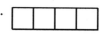   3c.

4a.    4b. 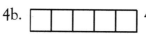   4c.

**Part C**

1.    2a. 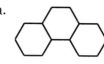   2b.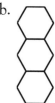

2c.    3a.    3b.

3c.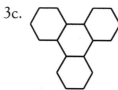

**Part D**

1.    2a.    2b.

3a.    3b.    4a.

4b.    4c.

Part E

1a.   1b.   1c.

1d.   2a. 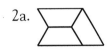  2b.

3a. 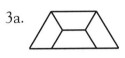  3b.   4a.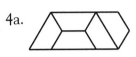

4b.

# INVESTIGATION 7

## LINES OF SYMMETRY

You need a set of pattern blocks, paper, and pencils.
    For each activity start with the given pattern block. Then, add the given number of pattern blocks so that the resulting figure has the given number of lines of symmetry. When blocks are added, entire sides of blocks that touch each other must match. Draw a picture of each pattern block figure that you make.

### Part A
Start with an orange square.

1. Add one green triangle to make a figure that has one line of symmetry.
2. Add two green triangles to make a figure that has
   a. no lines of symmetry
   b. one line of symmetry
   c. two lines of symmetry.
3. Add three green triangles to make a figure that has one line of symmetry.
4. Add four green triangles to make a figure that has four lines of symmetry.

### Part B
Start with an orange square.

1. Add one orange square to make a figure that has two lines of symmetry.
2. Add two orange squares to make a figure that has
   a. one line of symmetry
   b. two lines of symmetry
3. Add three orange squares to make a figure that has
   a. one line of symmetry
   b. two lines of symmetry
   c. four lines of symmetry
4. Add four orange squares to make a figure that has
   a. one line of symmetry
   b. two lines of symmetry
   c. four lines of symmetry

**Part C**

Start with a yellow hexagon.

1. Add one yellow hexagon to make a figure that has two lines of symmetry.
2. Add two yellow hexagons to make a figure that has
   a. one line of symmetry
   b. two lines of symmetry
   c. three lines of symmetry
3. Add three yellow hexagons to make a figure that has
   a. one line of symmetry
   b. two lines of symmetry
   c. three lines of symmetry

**Part D**

Start with a yellow hexagon.

1. Add one orange square to make a figure that has one line of symmetry.
2. Add two orange squares to make a figure that has
   a. one line of symmetry
   b. two lines of symmetry
3. Add three orange squares to make a figure that has
   a. one line of symmetry
   b. three lines of symmetry
4. Add four orange squares to make a figure that has
   a. no lines of symmetry
   b. one line of symmetry
   c. two lines of symmetry

## Part E

Start with a red trapezoid.

1. Add one red trapezoid to make a figure that has
   a. no lines of symmetry
   b. one line of symmetry
   c. two lines of symmetry
   d. six lines of symmetry
2. Add two red trapezoids to make a figure that has
   a. no lines of symmetry
   b. one line of symmetry
3. Add three red trapezoids to make a figure that has
   a. one line of symmetry
   b. two lines of symmetry
4. Add four red trapezoids to make a figure that has
   a. no lines of symmetry
   b. one line of symmetry

# INVESTIGATION 8

## FINDING CONGRUENT GEOBOARD FIGURES

This investigation requires students to use their understanding of the geometric concept of congruence and apply the geometric rigid motions of slides, flips, and turns. Students will have ample opportunities to exercise their spatial perception and visualization skills. Students need to have had previous experiences making polygons on geoboards.

### Student Materials
Each group needs 5-by-5 geoboards and rubber bands.

### Introduction of the Investigation
- Review the meaning of congruence and explain the procedure for each activity.
- Remind students that a figure is always congruent to itself; the total count of the number of figures congruent to it includes the given figure.

### Closure of the Investigation
The groups report their answers and explain their counting procedures. Compare and contrast the different methods of counting. Emphasize the need for a systematic counting procedure to ensure that all congruent figures have been counted. Point out how the following answers are related:

    A1   and   C1
    A3   and   C4
    B1   and   C3
    B1   and   D1
    B2   and   D3
    B3   and   D2

BASIC GEOMETRIC CONCEPTS

For example, in the square there are four congruent triangles.

So, the answer to C4 is 9 × 4 = 36, since in A3 there are nine such congruent squares on the geoboard.

**Answers**

**Part A**
1. 16
2. 4
3. 9

**Part B**
1. 24
2. 6
3. 12

**Part C**
1. 4
2. 64
3. 96
4. 36

**Part D**
1. 96
2. 24
3. 24
4. 32

# INVESTIGATION 8

## FINDING CONGRUENT GEOBOARD FIGURES

You need a 5-by-5 geoboard and rubber bands.

### Part A
Find how many squares on your geoboard are congruent to the given square. Include the given square in your count.

1.   2.   3.

### Part B
Find how many rectangles on your geoboard are congruent to the given rectangle. Include the given rectangle in your count.

1.   2.   3.

### Part C
Find how many right triangles on your geoboard are congruent to the given right triangle. Include the given right triangle in your count.

1.   2.   3.   4.

### Part D
Find how many trapezoids on your geoboard are congruent to the given trapezoid. Include the given trapezoid in your count.

1.   2.   3.   4.

BASIC GEOMETRIC CONCEPTS

# INVESTIGATION 9

## MAKING SIMILAR GEOMETRIC FIGURES

These activities are designed to help students understand the concept of similarity and discover the fundamental properties of similar geometric polygons. Students will also exercise their spatial perception and visualization skills and use the concept of ratio.

### Student Materials
Each group needs a set of pattern blocks.

### Introduction of the Investigation
- Outline the activities.
- Introduce the concept of similar geometric figures by showing examples.
- The groups are to discover for themselves the fundamental properties of similar polygons in Part F (corresponding angles are congruent and corresponding sides in are proportional).

### Closure of the Investigation
The groups report on their solutions to the activities. They may demonstrate their pattern block figures by making them on an overhead projector. In Part F, point out that the common ratio of corresponding sides of similar figures is called the *ratio of similarity* or the *scale factor*.

### Answers

#### Part A
1. Corresponding angles are congruent. (Students can place the green triangle on the pattern block triangles to show that corresponding angles coincide.)
2. $\frac{1}{2}, \frac{1}{3}, \frac{1}{4}$

#### Part B
1. Corresponding angles are congruent.
2. $\frac{1}{2}, \frac{1}{3}, \frac{1}{4}$

## Part C
a.

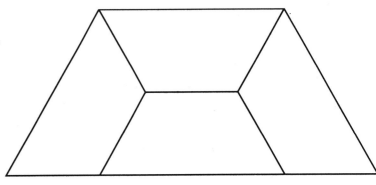

b.

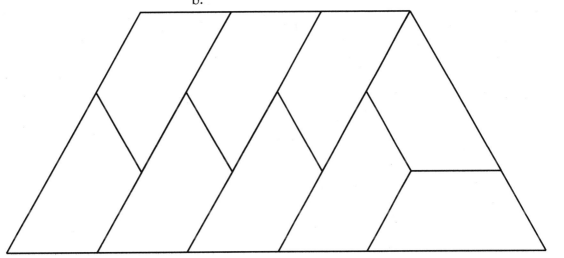

1. Corresponding angles are congruent.
2. a. $\frac{2}{4} = \frac{1}{2}$   b. $\frac{2}{6} = \frac{1}{3}$

## Part D
1. Corresponding angles are congruent.
2. a. $\frac{2}{4} = \frac{1}{2}$   b. $\frac{2}{6} = \frac{1}{3}$

## Part E
1. Corresponding angles are congruent.
2. $\frac{3}{2}$

## Part F
1. congruent
2. equal

# INVESTIGATION 9

## MAKING SIMILAR GEOMETRIC FIGURES

You need a set of pattern blocks.

Two geometric figures are *similar* if they have the same shape but not necessarily the same size. In the following activities, use your set of pattern blocks to make similar geometric figures and investigate the properties of similar geometric figures.

### Part A
Each side of the green triangle is one unit long. Use green triangles to make pattern block triangles similar to the green triangle having side lengths that are
   a. two units long
   b. three units long
   c. four units long

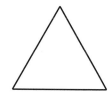

1. How do the angles of each pattern block triangle compare to the corresponding angles of the green triangle?

2. For each pattern block triangle, find the ratio of the side length of the green triangle to the side length of the pattern block triangle.

### Part B
Each side of the blue rhombus is one unit long. Use blue rhombuses to make pattern block rhombuses similar to the blue rhombus having side lengths that are
   a. two units long
   b. three units long
   c. four units long

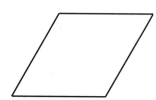

1. How do the angles of each pattern block rhombus compare to the corresponding angles of the blue rhombus?

2. For each pattern block rhombus, find the ratio of the side length of the blue rhombus to the side length of the pattern block rhombus.

## Part C

One side of the red trapezoid is two units long; each of its other sides is one unit long. Use red trapezoids to make pattern block trapezoids similar to the red trapezoid having side lengths that are

    a. four units and two units long
    b. six units and three units long

1. How do the angles of each pattern block trapezoid compare to the corresponding angles of the red trapezoid?

2. For each pattern block trapezoid, find the ratios of the two different side lengths of the red trapezoid to the two corresponding side lengths of the pattern block trapezoid.

## Part D

This parallelogram has sides that are two units long and one unit long. Use tan rhombuses to make pattern block parallelograms similar to this parallelogram having side lengths that are

    a. four units long and two units long
    b. six units long and three units long

1. How do the angles of each pattern block parallelogram compare to the corresponding angles of this parallelogram?

2. For each pattern block parallelogram, find the ratios of the two different side lengths of this parallelogram to the two corresponding side lengths of the pattern block parallelogram.

BASIC GEOMETRIC CONCEPTS

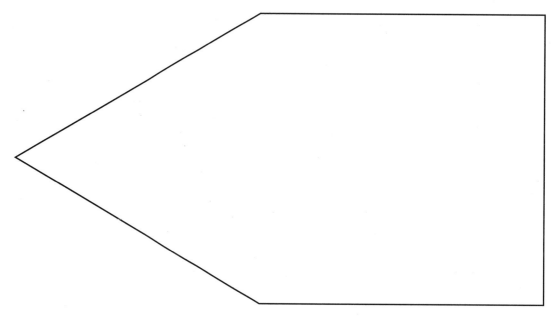

**Part E**

Each side of this pattern block pentagon is three units long. Use orange squares and green triangles to make a similar pattern block pentagon having sides that are each two units long.

1. How do the corresponding angles of the two pattern block pentagons compare?

2. Find the ratio of the side length of the larger pattern block pentagon to the side length of the smaller pattern block pentagon.

**Part F**

Use the results of your investigations in Parts A–E to complete the following statements.

1. If two geometric figures are similar, then their corresponding angles are _____ .
2. If two geometric figures are similar, then the ratios of their corresponding sides are _____ .

# ANGLES OF POLYGONS

**INVESTIGATION 10**     The Angles of a Triangle

**INVESTIGATION 11**     The Angles of a Quadrilateral

**INVESTIGATION 12**     The Angles of a Polygon

# INVESTIGATION 10

## THE ANGLES OF A TRIANGLE

This activity is designed for students to discover that the sum of the measures of the interior angles of a triangle is 180°. Students will also practice applying this important geometric property.

### Student Materials
Each group needs four triangular pieces of paper—one right triangle, one isosceles triangle, one acute scalene triangle, and one obtuse scalene triangle.

### Introduction of the Investigation
- Lead the class through Part A, demonstrating how to make the various folds on triangular region $ABC$. (Angles $A$, $B$, and $C$ should meet at point $D$ to form a straight line.)
- Each student in a group will make the folds on one triangle.

### Closure of the Investigation
Groups report and discuss their answers to the questions in Part B.

### Answers

#### Part B
1. 180°
2. 80°, 135°, 41°
3. 60°
4. 45°
5. The sum of the measures of the acute angles of a right triangle is 90°; that is, the two acute angles are complementary angles.
6. no, no, yes

# INVESTIGATION 10

## THE ANGLES OF A TRIANGLE

You need a collection of triangular pieces of paper.

### Part A

1. Choose a triangular region $ABC$.

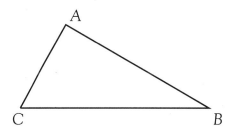

2. Fold vertex $C$ onto side $\overline{BC}$ so that when you make a crease, the crease will pass through vertex $A$.

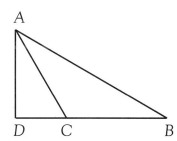

3. Unfold the paper. The crease $\overline{AD}$ that you made is the altitude from vertex $A$ to side $\overline{BC}$.

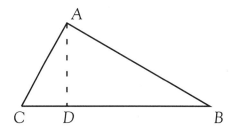

4. Fold vertex $A$ onto point $D$. Fold vertex $C$ onto point $D$. Fold vertex $B$ onto point $D$.
5. What is the sum of the measures of angles $A$, $B$, and $C$?

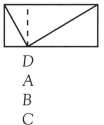

ANGLES OF POLYGONS

53

**Part B**

Use your result to question 1 below to answer the other questions in this part.

1. What is the sum of the measures of the angles of any triangle?

2. Find the measure of the unknown angle in each triangle.

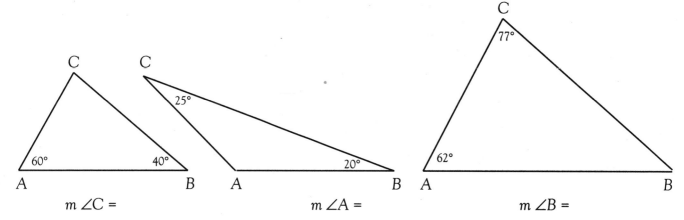

m ∠C =                    m ∠A =                    m ∠B =

3. What is the measure of each angle of an equilateral triangle?

4. What is the measure of each acute angle of an isosceles right triangle?

5. Name the relationship between the acute angles of any right triangle.

6. Can a triangle have exactly two right angles? Explain.

   Can a triangle have exactly two obtuse angles? Explain.

   Can a triangle have exactly two acute angles? Explain.

# INVESTIGATION 11

## THE ANGLES OF QUADRILATERAL

This activity is designed for students to learn and to apply the property that the sum of the measures of the interior angles of a convex quadrilateral is 360°.

### Introduction of the Investigation
- Explain the activity.
- In Part A students should not measure angles of quadrilaterals with protractors; rather, they should use the property discovered in Investigation 10.

### Closure of the Investigation
The groups discuss their answers to the questions in Part B.

### Answers

#### Part B
1. 65°, 53°
2. yes, yes, no, yes
3. yes, yes, yes, no
4. no
5. 210°, 97°, 180°

# INVESTIGATION 11

## THE ANGLES OF A QUADRILATERAL

### Part A
Without measuring angles, find the sum of the measures of the angles of each quadrilateral. (Hint: draw a diagonal of each quadrilateral.)

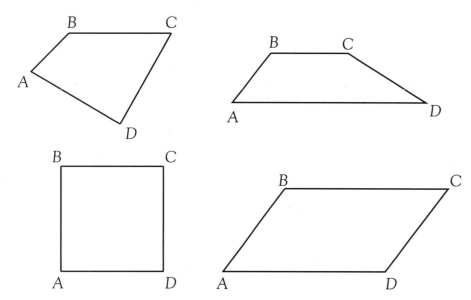

The sum of the measures of the angles of any quadrilateral is _____.

### Part B
Use your result in Part A to answer these questions in this part.

1. Find the measures of the unknown angle in each quadrilateral.

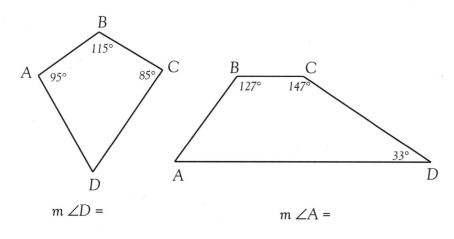

m ∠D =

m ∠A =

56          COOPERATIVE INFORMAL GEOMETRY

2. Can a quadrilateral have exactly one right angle? Explain.

   Can a quadrilateral have exactly two right angles? Explain.

   Can a quadrilateral have exactly three right angles? Explain.

   Can a quadrilateral have exactly four right angles? Explain.

3. Can a quadrilateral have exactly one obtuse angle? Explain.

   Can a quadrilateral have exactly two obtuse angles? Explain.

   Can a quadrilateral have exactly three obtuse angles? Explain.

   Can a quadrilateral have exactly four obtuse angles? Explain.

4. Can all of the angles of quadrilateral be acute angles? Explain.

5. Find each sum for the given quadrilateral.

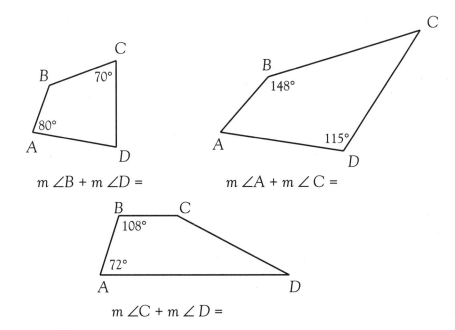

$m\angle B + m\angle D =$

$m\angle A + m\angle C =$

$m\angle C + m\angle D =$

ANGLES OF POLYGONS

# INVESTIGATION 12

## THE ANGLES OF A POLYGON

This activity is designed for students to discover the formula for the sum of the measures of the angles of a polygon.

**Introduction to the Investigation**
- Explain what the groups are to do in the activity.
- Students will need to use the property discovered in Investigation 10.

**Closure of the Investigation**
The groups discuss their answers to the activity.

**Answers**
1. 4, 2, 2 (180°) = 360°
   5, 3, 3 (180°) = 540°
   6, 4, 4 (180°) = 720°
   7, 5, 5 (180°) = 900°
   8, 6, 6 (180°) = 1080°
   9, 7, 7 (180°) = 1260°
   10, 8, 8 (180°) = 1440°
2. $(n-2)\,180°$
3. $90°, 108°, 120°, 128\frac{4}{7}°, 135°, 140°, 144°$
4. $\dfrac{(n-2)\,180°}{n}$

COOPERATIVE INFORMAL GEOMETRY

# INVESTIGATION 12

## THE ANGLES OF A POLYGON

1. Draw a picture of each polygon listed in the table. Choose one vertex of the polygon, draw all possible diagonals of the polygon from that vertex, and count the number of triangles that are formed. Then, complete the table.

| Polygon | Number of Sides | Number of Triangles | Sum of the Measures of the Angles of the Polygon |
|---|---|---|---|
| Quadrilateral | | | |
| Pentagon | | | |
| Hexagon | | | |
| Heptagon | | | |
| Octagon | | | |
| Nonagon | | | |
| Decagon | | | |

2. Determine the pattern that relates the number of sides of the polygon to the sum of the measures of the angles. Write a formula for the sum of the measures of the angles of a polygon with $n$ sides.

3. A polygon is *regular* if all sides are congruent and all angles are congruent. Use the information in your table in question 1 to find the measure of each angle of a regular polygon.

| Regular Polygon | Measure of Each Angle |
|---|---|
| Quadrilateral | |
| Pentagon | |
| Hexagon | |
| Heptagon | |
| Octagon | |
| Nonagon | |
| Decagon | |

4. Determine the pattern in question 3 that relates the number of sides of a regular polygon to the measure of each angle of the polygon. Write a formula for the measure of each angle of a regular polygon with *n* sides.

# PROPERTIES OF SPECIAL POLYGONS

**INVESTIGATION 13**  Properties of Isosceles Triangles

**INVESTIGATION 14**  Properties of Parallelograms

**INVESTIGATION 15**  Diagonals of Quadrilaterals

# INVESTIGATION 13

## PROPERTIES OF ISOSCELES TRIANGLES

This paper-folding activity leads students to discover several special properties of isosceles triangles.

### Student Materials
Each group needs a variety of paper isosceles triangles including acute, obtuse, and right triangles.

### Introduction of the Investigation
- Introduce the class to Part A by defining an isosceles triangle.
- Demonstrate how to fold the bisector of angle C.

### Closure of the Investigation
The groups report and discuss what they have discovered in Parts A and B.

### Answers

#### Part A
1. The angles opposite the congruent sides are congruent.
2. The bisector of the angle formed by the two congruent sides bisects the opposite side of the triangle and is also perpendicular to that side (since it forms two congruent angles that are supplementary).
3. The bisector of the angle formed by the two congruent sides is the only line of symmetry of the isosceles triangle.

#### Part B
1. $m \angle A + m \angle B = 90°$
2. $m \angle A = m \angle B = 45°$

#### Part C
Triangle ABC is isosceles, since $\overline{CA}$ and $\overline{CB}$ have the same length.

# INVESTIGATION 13

## PROPERTIES OF ISOSCELES TRIANGLES

You need a collection of different isosceles triangles, a sheet of paper, and a centimeter ruler.

### Part A

An *isosceles triangle* is a triangle with two congruent sides. To discover the properties of isosceles triangles, fold the bisector of the angle formed by the two congruent sides of the triangle for each triangle in your collection.

If sides $\overline{AC}$ and $\overline{BC}$ are congruent, fold side $\overline{AC}$ onto side $\overline{BC}$ so that point A falls on point B. Make the crease $\overline{CD}$.

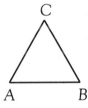 

Unfold the paper to get back triangle ABC.

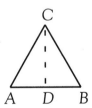

Repeat this for all the isosceles triangles in your collection.

1. What is true about the angles of an isosceles triangle?

2. What is true about the bisector of the angle formed by the two congruent sides and the side opposite that angle?

3. Does an isosceles triangle have any lines of symmetry?

PROPERTIES OF SPECIAL POLYGONS

## Part B

For each isosceles right triangle in your collection, fold vertex A onto vertex C and make a crease; then, fold vertex B onto vertex C and make a crease.

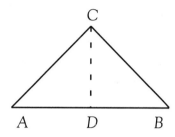 

What is true about $m \angle A + m \angle B$?

Find $m \angle A$ and $m \angle B$.

## Part C

On a sheet of paper mark two points A and B, and draw line segment $\overline{AB}$. Fold the paper so that point A falls on point B, pinch and crease the paper, then unfold the paper.

The line you folded is the perpendicular bisector of line segment $\overline{AB}$. Choose any point C on that line. What kind of triangle is ABC? Explain.

# INVESTIGATION 14

## PROPERTIES OF PARALLELOGRAMS

This activity is designed to help students discover the common properties of parallelograms. Geometric concepts involved include congruence, bisection point, and supplementary angles.

### Student Materials
Each student needs a copy of the sheet of geoboard parallelograms of different sizes and shapes, a centimeter ruler, and a protractor.

### Introduction of the Investigation
- Explain the activity.
- Mention that each geoboard figure in the collection is an example of a parallelogram.

### Closure of the Investigation
The groups report their discoveries about the common properties of a parallelogram. Emphasize the fact that rectangles, squares, and rhombuses are special kinds of parallelograms.

### Answers
1. The opposite sides of any parallelogram are congruent.
2. The opposite angles of any parallelogram are congruent.
3. The consecutive angles of any parallelogram are supplementary. A parallelogram, therefore, has four pairs of supplementary angles.
4. The diagonals of any parallelogram bisect each other. (If the parallelogram is a rectangle, then its diagonals are congruent. If the parallelogram is a rhombus, then its diagonals are perpendicular.)

# GEOBOARD PARALLELOGRAMS

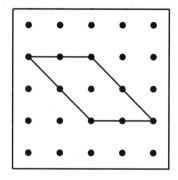 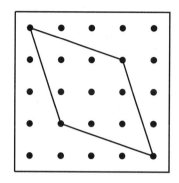 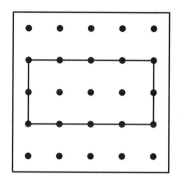

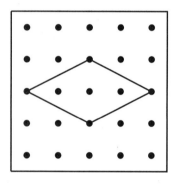 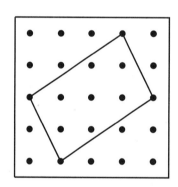 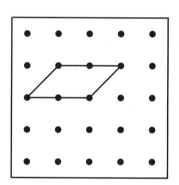

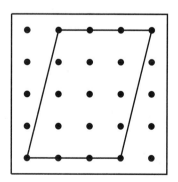 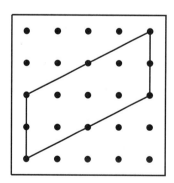 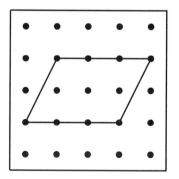

# INVESTIGATION 14

## PROPERTIES OF PARALLELOGRAMS

You need a collection of geoboard parallelograms of different sizes and shapes, a centimeter ruler, and a protractor.

A *parallelogram* is a quadrilateral with two pairs of parallel sides. For each parallelogram in your collection, make the following measurements and write the measurements on the figure.

    a. Measure each side of the parallelogram to the nearest tenth of a centimeter.

    b. Measure each angle of the parallelogram to the nearest degree.

    c. Draw the diagonals of the parallelogram and measure the diagonals to the nearest tenth of a centimeter.

Look at your collection of parallelograms and their measurements.

1. What is true about the sides of any parallelogram?

2. What is true about the angles of any parallelogram?

3. Remember that two angles are *supplementary* if the sum of their measures is 180°. Does a parallelogram have any pairs of supplementary angles? If it does, which ones are supplementary?

4. What is true about the diagonals of any parallelogram? How are the diagonals related to each other?

PROPERTIES OF SPECIAL POLYGONS

# INVESTIGATION 15

## THE DIAGONALS OF QUADRILATERALS

This activity is designed to help students discover the properties of the diagonals of parallelograms, rectangles, squares, rhombuses, and trapezoids.

### Student Materials
Each group needs copies of two sheets of geoboard quadrilaterals, centimeter rulers, and protractors.

### Introduction of the Investigation
- Explain the activity, reviewing, if necessary, key concepts such as congruence, perpendicularity, the bisector of an angle, and the bisection point of a line segment.
- Students understand that if they write "yes" for an entry in the table, it means they have decided that the diagonal property is true for all quadrilaterals of that category. Students may need to measure line segments to the nearest tenth of a centimeter and to measure angles to the nearest degree to check for congruence.

### Closure of the Investigation
The groups report their discoveries regarding the diagonals of the various quadrilaterals. The class should reach consensus for each entry in the table.

### Answers
Parallelogram: no, no, yes, no
Rectangle: yes, no, yes, no
Square: yes, yes, yes, yes
Rhombus: no, yes, yes, yes
Trapezoid: no, no, no, no

# GEOBOARD QUADRILATERALS (1)

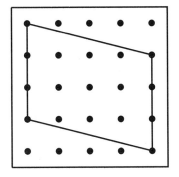 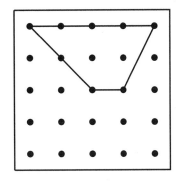 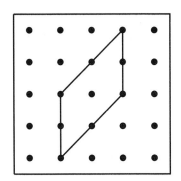

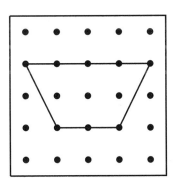 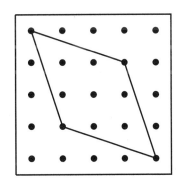 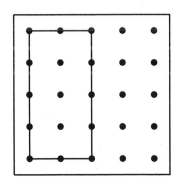

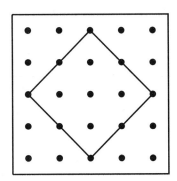 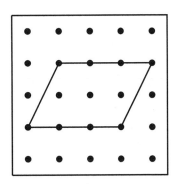 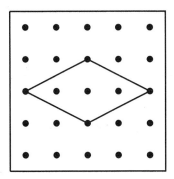

PROPERTIES OF SPECIAL POLYGONS

# GEOBOARD QUADRILATERALS (2)

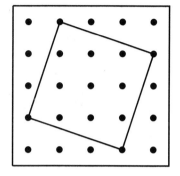
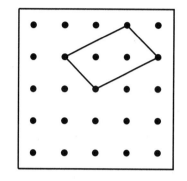
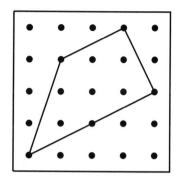
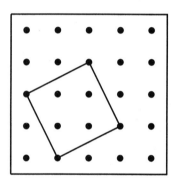
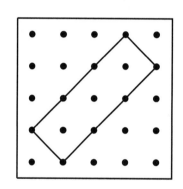
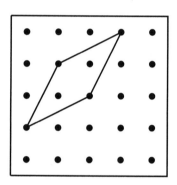
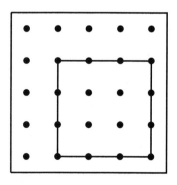
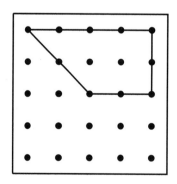
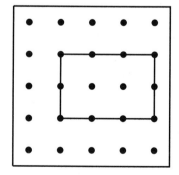

# INVESTIGATION 15

## THE DIAGONALS OF QUADRILATERALS

You need two sheets of geoboard quadrilaterals of different kinds (parallelograms, rectangles, squares, rhombuses, trapezoids), a centimeter ruler, and a protractor.

For each geoboard quadrilateral, name the quadrilateral. Then draw its diagonals and investigate the properties of its diagonals. Are they congruent? Are they perpendicular? Do they bisect each other? Do they bisect the angles whose vertices they join? Record the results of your investigations by completing the table by writing "yes" or "no" in each box.

| Quadrilateral | Diagonals Are Congruent | Diagonals Are Perpendicular | Diagonals Bisect Each Other | Diagonals Bisect Angles |
|---|---|---|---|---|
| Parallelogram | | | | |
| Rectangle | | | | |
| Square | | | | |
| Rhombus | | | | |
| Trapezoid | | | | |

PROPERTIES OF SPECIAL POLYGONS

# AREA AND PERIMETER

**INVESTIGATION 16**     Pattern Block Areas

**INVESTIGATION 17**     Areas of Geoboard Squares and Rectangles

**INVESTIGATION 18**     Areas of Geoboard Right Triangles

**INVESTIGATION 19**     Finding the Largest and Smallest Areas

**INVESTIGATION 20**     Increasing the Perimeter

# INVESTIGATION 16

## PATTERN BLOCK AREAS

This investigation teaches the meaning of the concept of area by using nonstandard units of area.

### Student Materials
Each group needs a set of pattern blocks.

### Introduction of the Investigation
- Discuss with the class the meaning of area: to find the area of a plane geometric shape, first choose a geometric shape, define it to have an area of one unit, and count how many copies of that unit shape are needed to "fill up" or cover the given geometric shape. Area, then, is a measure of covering.
- Explain each activity, indicating that some answers will be fractions or mixed numbers.

### Closure of the Investigation
The groups explain their answers to the activities. They may use an overhead projector to demonstrate their solutions. Students must understand that the numerical value of the area of a geometric shape depends upon the choice of the unit of area.

### Answers
1. 2, 3, 6
2. 1, 2, 3, 6

   $\frac{1}{2}, 1, 1\frac{1}{2}, 3$

   $\frac{1}{3}, \frac{2}{3}, 1, 2$

   $\frac{1}{6}, \frac{1}{3}, \frac{1}{2}, 1$

3. a. 18, 16, 12, 20

   b. 9, 8, 6, 10

   c. 6, $5\frac{1}{3}$, 4, $6\frac{2}{3}$

   d. 3, $2\frac{2}{3}$, 2, $3\frac{1}{3}$

**Extension**

Ask the students to look for a pattern in their answers to questions 2 and 3. In question 2, how are the answers in the other rows of the table related to the answers in the first row? In question 3, how are the answers to Parts B–D related to the answers to Part A?

# INVESTIGATION 16

## PATTERN BLOCK AREAS

You need a set of pattern blocks and copies of Shapes A–D.

1. If the area of the green triangle is one unit, find the area of the following pattern blocks:
   blue rhombus:     area = _____ units
   red trapezoid:    area = _____ units
   yellow hexagon:   area = _____ units

2. Complete the following table:

|      | Green Triangle | Blue Rhombus | Red Trapezoid | Yellow Hexagon |
|------|----------------|--------------|---------------|----------------|
| Area | one unit       |              |               |                |
| Area |                | one unit     |               |                |
| Area |                |              | one unit      |                |
| Area |                |              |               | one unit       |

3. Find the areas of Shapes A, B, C, and D
   a. if the green triangle has an area of one unit.

   A _____ B _____ C _____ D _____

   b. if the blue rhombus has an area of one unit.

   A _____ B _____ C _____ D _____

   c. if the red trapezoid has an area of one unit.

   A _____ B _____ C _____ D _____

   d. if the yellow hexagon has an area of one unit.

   A _____ B _____ C _____ D _____

AREA AND PERIMETER

Shape A

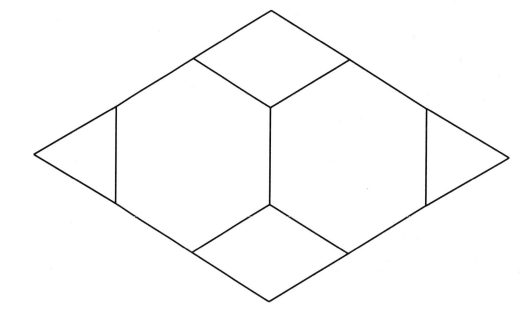

Shape B

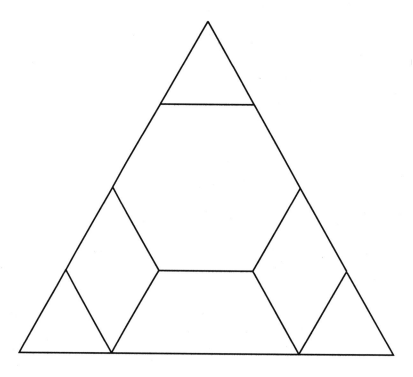

Shape C

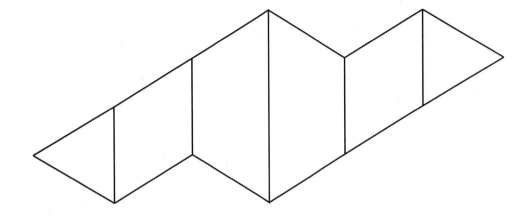

Shape D

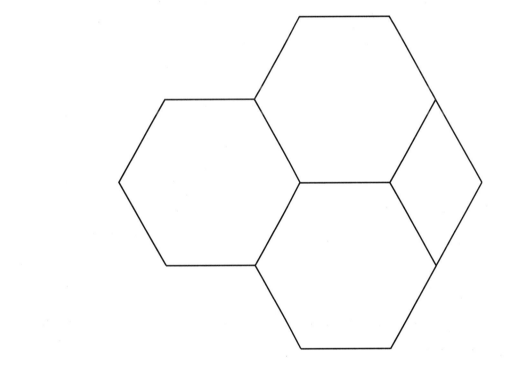

# INVESTIGATION 17

## AREAS OF GEOBOARD SQUARES AND RECTANGLES

This investigation teaches the meaning of area of squares and rectangles as the number of unit squares needed to cover or "fill up" the given figure. Students will apply their understanding of squares and rectangles and will use their spatial perception and visualization skills. Students need to have had previous experiences with geoboards.

### Student Materials
Each group needs 5-by-5 geoboards, rubber bands, and geoboard dot paper (see Investigation 1).

### Introduction of the Investigation
- Discuss the meaning of area as the number of unit squares needed to cover or "fill up" a given geometric figure.
- Explain the activity. You may choose to discuss a few sample activities with the entire class, especially those where fractions of unit squares may be needed in finding areas.

### Closure of the Investigation
The groups explain their answers. A transparent geoboard for the overhead projector may be useful.

### Answers

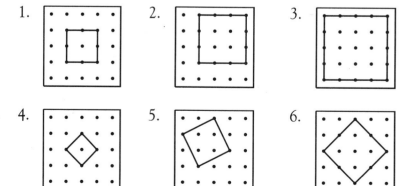

COOPERATIVE INFORMAL GEOMETRY

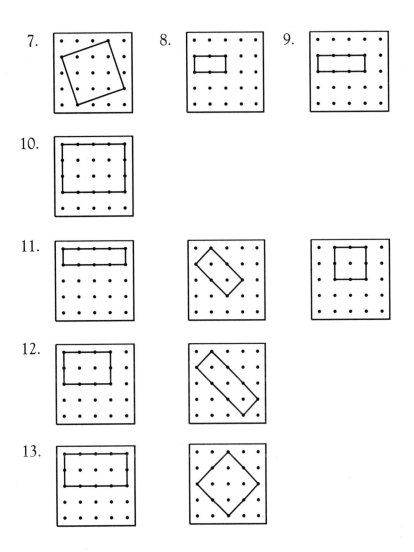

**Extension**

For the geoboard squares and rectangles having sides that are horizontal and vertical, have students to count the number of rows of unit squares and the number of unit squares per row. Have them relate this data to the area of the figure, and ask for the generalization. This, of course, leads to the familiar area formulas $A = s \times s$ and $A = l \times w$.

AREA AND PERIMETER

# INVESTIGATION 17

## AREAS OF GEOBOARD SQUARES AND RECTANGLES

You need a 5-by-5 geoboard, rubber bands, and geoboard dot paper.
   Let the geoboard square ☐ have an area of one square unit. Create the following geoboard squares or rectangles with the given areas; record your geoboard figures on geoboard dot paper.

1. A square with an area of four square units.
2. A square with an area of nine square units.
3. A square with an area of sixteen square units.
4. A square with an area of two square units.
5. A square with an area of five square units.
6. A square with an area of eight square units.
7. A square with an area of ten square units.
8. A nonsquare rectangle with an area of two square units.
9. A rectangle with an area of three square units.
10. A rectangle with an area of twelve square units.
11. Two noncongruent rectangles, each with an area of four square units.
12. Two noncongruent rectangles, each with an area of six square units.
13. Two noncongruent rectangles, each with an area of eight square units.

# INVESTIGATION 18

## AREAS OF GEOBOARD RIGHT TRIANGLES

This investigation teaches the meaning of area of a right triangle as the number of unit squares needed to cover or "fill up" the right triangle. Students will apply their understanding of right triangles and use their spatial perception and visualization skills. Students need to have had previous experiences with geoboards.

### Student Materials
Each group needs 5-by-5 geoboards, rubber bands, and geoboard dot paper (see Investigation 1).

### Introduction of the Investigation
- Discuss the meaning of area as the number of unit squares needed to cover or "fill up" a given geometric figure.
- Explain what the groups are to do in each activity.
- You may choose to discuss a few of the activities with the entire class as sample activities, pointing out strategies for finding the area of a right triangle by viewing it as one-half of a rectangle or by breaking it up into smaller figures with known areas.

### Closure of the Investigation
The groups explain their answers. A transparent geoboard for the overhead projector may be useful.

### Answers

1. 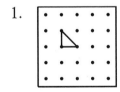  2.   3.

4.   5.   6.

AREA AND PERIMETER

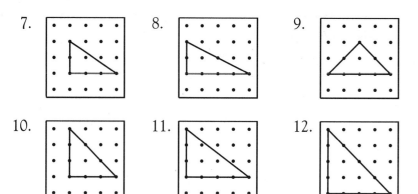

**Extension**

Discuss the strategy of enclosing a right triangle with base $b$ and height $h$ in a rectangle with base $b$ and height $h$, and lead students to the discovery of the formula for finding the area of a triangle, $A = \frac{1}{2}bh$.

# INVESTIGATION 18

## AREAS OF GEOBOARD RIGHT TRIANGLES

You need a 5-by-5 geoboard, rubber bands, and geoboard dot paper.
  Let the geoboard square ☐ have an area of one square unit. Create the following geoboard right triangles with the given descriptions. Record your geoboard right triangles on geoboard dot paper.

1. A right triangle with an area of $\frac{1}{2}$ square unit

2. A right triangle with an area of one square unit

3. A right triangle with an area of one square unit that is not congruent to your right triangle in question 2

4. A right triangle with an area of $1\frac{1}{2}$ square units

5. A right triangle with an area of two square units

6. A right triangle with an area of two square units that is not congruent to your right triangle in question 5

7. A right triangle with an area of three square units

8. A right triangle with an area of four square units

9. A right triangle with an area of four square units that is not congruent to your right triangle in question 8

10. A right triangle with an area of $4\frac{1}{2}$ square units

11. A right triangle with an area of six square units

12. A right triangle with an area of eight square units

# INVESTIGATION 19

## FINDING THE LARGEST AND SMALLEST AREAS

In this discovery-learning, problem-solving activity students confront the meanings of area and perimeter and the differences between them. Students will also use their skills of spatial perception and gain experience in looking for patterns.

### Student Materials
Each group needs a set of 40 color tiles.

### Introduction of the Investigation
- Review the definitions of area and perimeter and the difference in what they are measuring.
- Outline the different parts of the investigation.
- Remind the groups that they are looking for all noncongruent rectangles with the same perimeter. Also encourage the groups to search for patterns as they work through the activities in Part A.

### Closure of the Investigation
The groups discuss their answers.

### Answers

Part A

1. 5, 1, 12, 5
   4, 2, 12, 8
   3, 3, 12, 9

2. 6, 1, 14, 6
   5, 2, 14, 10
   4, 3, 14, 12

3. 7, 1, 16, 7
   6, 2, 16, 12
   5, 3, 16, 15
   4, 4, 16, 16

4. 8, 1, 18, 8
   7, 2, 18, 14
   6, 3, 18, 18
   5, 4, 18, 20
5. 9, 1, 20, 9
   8, 2, 20, 16
   7, 3, 20, 21
   6, 4, 20, 24
   5, 5, 20, 25
6. 10, 1, 22, 10
   9, 2, 22, 18
   8, 3, 22, 24
   7, 4, 22, 28
   6, 5, 22, 30
7. 11, 1, 24, 11
   10, 2, 24, 20
   9, 3, 24, 27
   8, 4, 24, 32
   7, 5, 24, 35
   6, 6, 24, 36

**Part B**

1. The rectangle having the largest area is the square or the rectangle that is closest in shape to a square.
2. The rectangle having the smallest area is the rectangle that is the narrowest in shape. (It is the rectangle that has one unit for one of its dimensions.)
3. $p = 2 \times (l + w)$, so $l + w = \frac{p}{2}$

   Find all pairs of positive integers $l$ and $w$ whose sum is $\frac{p}{2}$, disregarding order of the integers, since rectangles with dimensions $l$ by $w$ and $w$ by $l$ are congruent.
4. 49, 13
5. 56, 14

# INVESTIGATION 19

## FINDING THE LARGEST AND SMALLEST AREAS

You need a set of 40 color tiles.

### Part A

1. Use your set of color tiles to find all *noncongruent* rectangles having a perimeter of 12 units. Find the length and width of each rectangle, its perimeter, and its area.

2. Use your set of color tiles to find all *noncongruent* rectangles having a perimeter of 14 units. Find the length and width of each rectangle, its perimeter, and its area.

3. Repeat the activity using a perimeter of 16 units.

4. Repeat the activity using a perimeter of 18 units.

5. Repeat the activity using a perimeter of 20 units.

6. Repeat the activity using a perimeter of 22 units.

7. Repeat the activity using a perimeter of 24 units.

**Part B**
Study your results in Part A and look for patterns.

1. In each case describe the kind of rectangle that has the largest area.

2. In each case describe the kind of rectangle that has the smallest area.

3. Describe how to find all noncongruent rectangles, without using color tiles, whose lengths and width are integers and whose perimeters are the same.

4. If the perimeter of a rectangle is 28 units and the length and width are integers, predict the area of the rectangle with the largest area and the area of the rectangle with the smallest area.

5. If the perimeter is 30 units and the length and width are integers, predict the area of the rectangle with the largest area and the area of the rectangle with the smallest area.

# INVESTIGATION 20

## INCREASING THE PERIMETER

In this discovery-learning, problem-solving activity students confront the meanings of area and perimeter and the differences between them. Students will also use their skills of spatial perception and gain experience in looking for patterns.

### Student Materials
Each group needs a set of about 30 color tiles of different colors.

### Introduction of the Investigation
- Explain how to form shapes with color tiles emphasizing that at least one whole side of a tile must touch one whole side of another tile.
- Review the meanings of area and perimeter and the differences between them. Area is a measure of the covering of a region with square units; to find the area of a color-tile shape, count the number of tiles. Perimeter is a measure of the distance around a figure; to find the perimeter of a color-tile shape, count the number of sides of the tiles on the outside of the shape.
- Discuss Part A so that students will understand how the perimeter of a shape can change when just one tile is added. This information is very useful in Parts B and C.

### Closure of the Investigation
The groups discuss their answers to Parts B–E.

### Answers

Part A
1. increases by two units
2. remains the same
3. decreases by two units

Part B
2. 2, 7
3. 11, 16

### Part C
1. 10, 3, 11
2. 12, 2, 11
3. 12, 2, 11

### Part D
The final shape in each case is a 4 × 4 square with area 16 square units.

### Part E
The perimeters are always even numbers. Adding a tile to a shape either increases or decreases the perimeter by two units or does not change the perimeter. So the perimeter will still be an even number, since an even number plus two or an even number minus two is still an even number.

# INVESTIGATION 20

## INCREASING THE PERIMETER

You need a set of about 30 color tiles with different colors.

### Part A
The shape shown here has an area of six square units and a perimeter of 14 units.

Build this shape using only one color. Now, using other colors, add one tile to the shape. A tile must be added so that at least one whole side of the tile touches one whole side of another tile. There are three different effects on the perimeter of the shape that are possible when you add just one tile. They are

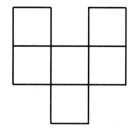

1. When one pair of sides match, the perimeter
   _____.

2. When two pairs of sides match, the perimeter
   _____.

3. When three pairs of sides match, the perimeter
   _____.

### Part B
The shape shown here has an area of five square units and a perimeter of 12 units.

Build this shape using only one color. Now, using other colors, add tiles so that the perimeter increases to 16 units. Tiles must be added so that at least one whole side of a tile touches one whole side of another tile.

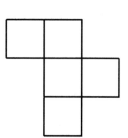

1. Find different ways to do this. Count how many tiles you added. Find the perimeter and the area of each new shape.
2. What is the fewest number of tiles that can be added to increase the perimeter to 16 units?
   What is the area of this new shape?
3. What is the greatest number of tiles that can be added to increase the perimeter to 16 units?
   What is the area of this new shape?

**Part C**

The perimeter of the five tiles in Part B may change if they are arranged in a different shape.

1. Arrange the five tiles in this shape.
   Its perimeter is _____ .
   Find the fewest number of tiles and the greatest number of tiles that can be added to increase the perimeter to 16 units.
   Fewest number _____
   Greatest number _____

2. Arrange the five tiles in this shape. Its perimeter is _____ . Find the fewest number of tiles and the greatest number of tiles that can be added to increase the perimeter to 16 units.
   Fewest number _____
   Greatest number _____

3. Arrange the five tiles in this shape. Its perimeter is _____ . Find the fewest number of tiles and the greatest number of tiles that can be added to increase the perimeter to 16 units.
   Fewest number _____
   Greatest number _____

**Part D**

What do you notice in each case about the final shape you get after adding the greatest number of tiles to increase the perimeter to 16? What is its area?

**Part E**

The perimeters of shapes made from color tiles that join only along complete sides always have something in common. What is it? Explain why this is true.

AREA AND PERIMETER

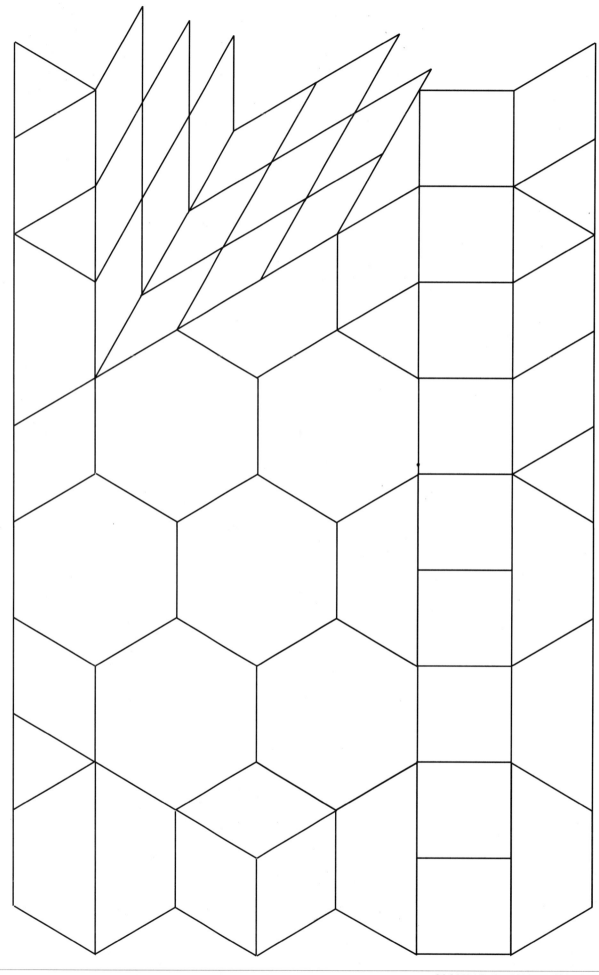